Chapter 6
Subtraction

W9-AXM-093

Houghton
Mifflin
Harcourt

© Houghton Mifflin Harcourt Publishing Company • Cover Image Credits: (Hares) ©Radius Images/Corbis; (Garden, New York) ©Rick Lew/The Image Bank/Getty Images; (sky) ©PhotoDisc/Getty Images

 Made in the United States
Text printed on 100%
recycled paper

Houghton Mifflin Harcourt

Curious George by Margret and H.A. Rey. Copyright © 2010 by Houghton Mifflin Harcourt Publishing Company. All rights reserved. The character Curious George®, including without limitation the character's name and the character's likenesses, are registered trademarks of Houghton Mifflin Harcourt Publishing Company.

Copyright © 2015 by Houghton Mifflin Harcourt Publishing Company

All rights reserved. No part of this work may be reproduced or transmitted in any form or by any means, electronic or mechanical, including photocopying or recording, or by any information storage and retrieval system, without the prior written permission of the copyright owner unless such copying is expressly permitted by federal copyright law. Requests for permission to make copies of any part of the work should be addressed to Houghton Mifflin Harcourt Publishing Company, Attn: Contracts, Copyrights, and Licensing, 9400 Southpark Center Loop, Orlando, Florida 32819-8647.

Common Core State Standards © Copyright 2010. National Governors Association Center for Best Practices and Council of Chief State School Offi cers. All rights reserved.

This product is not sponsored or endorsed by the Common Core State Standards Initiative of the National Governors Association Center for Best Practices and the Council of Chief State School Offi cers.

Printed in the U.S.A.

ISBN 978-0-544-34174-6

20 0928 20

4500802882   C D E F G

If you have received these materials as examination copies free of charge, Houghton Mifflin Harcourt Publishing Company retains title to the materials and they may not be resold. Resale of examination copies is strictly prohibited.

Possession of this publication in print format does not entitle users to convert this publication, or any portion of it, into electronic format.

Dear Students and Families,

Welcome to **Go Math!**, Grade K! In this exciting mathematics program, there are hands-on activities to do and real-world problems to solve. Best of all, you will write your ideas and answers right in your book. In **Go Math!**, writing and drawing on the pages helps you think deeply about what you are learning, and you will really understand math!

By the way, all of the pages in your **Go Math!** book are made using recycled paper. We wanted you to know that you can Go Green with **Go Math!**

Sincerely,

The Authors

Made in the United States
Text printed on 100% recycled paper

© Houghton Mifflin Harcourt Publishing Company • Image Credits: (bg) ©Sankar Salvady/Flickr/Getty Images; (t) ©Blaine Harrington II /Alamy Images; (c) ©Don Johnston/All Canada Photos/Getty Images; (b) ©Erich Kuchling/Westend61/Corbis

GO MATH!

Authors

Juli K. Dixon, Ph.D.
Professor, Mathematics Education
University of Central Florida
Orlando, Florida

Edward B. Burger, Ph.D.
President, Southwestern University
Georgetown, Texas

Steven J. Leinwand
Principal Research Analyst
American Institutes for
  Research (AIR)
Washington, D.C.

Contributor

Rena Petrello
Professor, Mathematics
Moorpark College
Moorpark, California

Matthew R. Larson, Ph.D.
K-12 Curriculum Specialist for
  Mathematics
Lincoln Public Schools
Lincoln, Nebraska

Martha E. Sandoval-Martinez
Math Instructor
El Camino College
Torrance, California

English Language Learners Consultant

Elizabeth Jiménez
CEO, GEMAS Consulting
Professional Expert on English
  Learner Education
Bilingual Education and
  Dual Language
Pomona, California

© Houghton Mifflin Harcourt Publishing Company • Image Credits: (bg) ©Russ Bishop/Alamy Images; (t) ©Richard Wear/Design Pics/Corbis

Number and Operations

 Common Core

Critical Area Representing, relating, and operating on whole numbers, initially with sets of objects.

GO DIGITAL

Go online! Your math lessons are interactive. Use *iTools*, Animated Math Models, the Multimedia *e*Glossary, and more.

6 **Subtraction** **307**

COMMON CORE STATE STANDARDS
K.OA Operations and Algebraic Thinking
Cluster A Understand addition as putting together and adding to, and understand subtraction as taking apart and taking from.
K.OA.A.1, K.OA.A.2, K.OA.A.5

✓ Show What You Know 308

Vocabulary Builder 309

Game: Spin for More 310

Chapter Vocabulary Cards

Vocabulary Game 310A

I Subtraction: Take From 311
Practice and Homework

2 Hands On • Subtraction: Take Apart 317
Practice and Homework

3 Problem Solving • Act Out Subtraction Problems 323
Practice and Homework

4 Hands On: **Algebra** • Model and Draw
Subtraction Problems 329
Practice and Homework

✓ Mid-Chapter Checkpoint 332

Chapter 6 Overview

In this chapter, you will explore and discover answers to the following **Essential Questions**:

• How can you show subtraction?
• How can you use numbers and symbols to show a subtraction sentence?
• How can using objects and drawings help you solve word problems?
• How can acting it out help you solve subtraction word problems?
• How can using addition help you solve subtraction word problems?

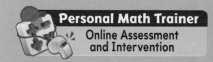

Personal Math Trainer
Online Assessment and Intervention

© Houghton Mifflin Harcourt Publishing Company

5 Algebra • Write Subtraction Sentences **335**
   Practice and Homework

6 Algebra • Write More Subtraction Sentences **341**
   Practice and Homework

7 Hands On: Algebra • Addition and Subtraction **347**
   Practice and Homework

✓ **Chapter 6 Review/Test** **353**

FOR MORE PRACTICE
GO TO THE
Personal Math Trainer

Practice and Homework

Lesson Check and
Spiral Review in
every lesson

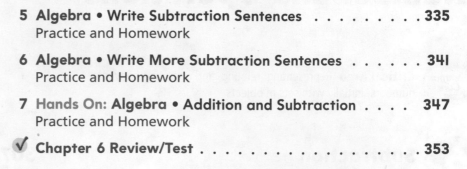

© Houghton Mifflin Harcourt Publishing Company

6 Subtraction

© Houghton Mifflin Harcourt Publishing Company • Image Credits: ©Randy Wells/Corbis
Curious George by Margret and H.A. Rey. Copyright © 2010 by Houghton Mifflin Harcourt Publishing Company.
All rights reserved. The character Curious George®, including without limitation the character's name and the
character's likenesses, are registered trademarks of Houghton Mifflin Harcourt Publishing Company.

Curious About Math with

Curious George

Penguins are birds with black and white feathers.

- There are 4 penguins on the ice. One penguin jumps in the water. How many penguins are on the ice now?

Name _____

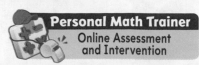
Fewer

1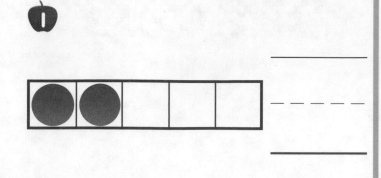

2

3

This page checks understanding of important skills needed for success in Chapter 6.

DIRECTIONS 1–2. Count and tell how many. Draw a set with one fewer counter. Write how many in each set. **3.** Write the number of cubes in each set. Circle the number that is less than the other number.

© Houghton Mifflin Harcourt Publishing Company

Name _____

add

© Houghton Mifflin Harcourt Publishing Company

DIRECTIONS Add the set of bees and the set of butterflies. Write how many insects altogether.

• **Interactive Student Edition**
• **Multimedia eGlossary**

Game

Spin for More

Spin for More				
Player 1				
Player 2				

DIRECTIONS Play with a partner. Decide who goes first. Take turns spinning to get a number from each spinner. Use cubes to model a cube train with the number from the first spin. Say the number. Add the cubes from the second spin. Compare your number with your partner's. Mark an X on the table for the player who has the greater number. The first player to have five Xs wins the game.

MATERIALS two paper clips, connecting cubes

© Houghton Mifflin Harcourt Publishing Company

Chapter 6 Vocabulary

add

sumar

2

fewer

menos

23

is equal to

es igual a

36

minus (−)

menos (−)

42

pairs

pares

50

plus (+)

más (+)

51

subtract

restar

72

zero

cero, ninguno

86

© Houghton Mifflin Harcourt Publishing Company

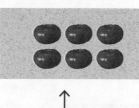

3 **fewer** birds

© Houghton Mifflin Harcourt Publishing Company

$2 + 4 = 6$

© Houghton Mifflin Harcourt Publishing Company

$6 - 3 = 3$

↙ **minus** sign

© Houghton Mifflin Harcourt Publishing Company

$3 + 2 = 5$

→ **is equal to**

© Houghton Mifflin Harcourt Publishing Company

$2 + 2 = 4$

↖ **plus** sign

© Houghton Mifflin Harcourt Publishing Company

3

3	0
2	1
1	2
0	3

pairs for 3

© Houghton Mifflin Harcourt Publishing Company
Credit: ©Artville/Getty Images

six tomatoes   zero tomatoes

© Houghton Mifflin Harcourt Publishing Company

$5 - 2 = 3$

Picture It

© Houghton Mifflin Harcourt Publishing Company • Image Credits: ©pablo_hernan/Fotolia

Word Box

minus
subtract
is equal to
plus
add
fewer
zero
pair

Secret Words

Player 1				
Player 2				

DIRECTIONS Players take turns. A player chooses a secret word from the Word Box and then sets the timer. The player draws pictures to give hints about the secret word. If the other player guesses the secret word before time runs out, he or she puts a counter in the chart. The first player who has counters in all his or her boxes is the winner.

MATERIALS timer, drawing paper, two-color counters for each player

The Write Way

DIRECTIONS Draw to show how to solve a subtraction problem. Write a subtraction sentence.
Reflect Be ready to tell about your drawing.

© Houghton Mifflin Harcourt Publishing Company • Image Credits: ©pablo_hernan/Fotolia

Name _____

Subtraction: Take From

Essential Question How can you show subtraction as taking from?

Common Core **Operations and Algebraic Thinking—K.OA.A.1**
MATHEMATICAL PRACTICES
MP1, MP2

Listen and Draw Real World

 take away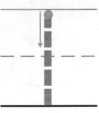

DIRECTIONS Listen to the subtraction word problem. Trace the number that shows how many children in all. Trace the number that shows how many children are leaving. Trace the number that shows how many children are left.

Chapter 6 • Lesson 1

three hundred eleven **311**

© Houghton Mifflin Harcourt Publishing Company

1

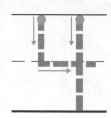 take away _____

_____

_____

- - - - -

_____

DIRECTIONS 1. Listen to the subtraction word problem. Trace the number that shows how many children in all. Write the number that shows how many children are leaving. Write the number that shows how many children are left.

© Houghton Mifflin Harcourt Publishing Company

Name _____

_____   _____
- - - - -   - - - - -

_____   take away   _____

_____
- - - - -

_____

© Houghton Mifflin Harcourt Publishing Company

DIRECTIONS **2.** Listen to the subtraction word problem. Write the number that shows how many children in all. Write the number that shows how many children are leaving. Write the number that shows how many children are left.

Chapter 6 • Lesson 1

three hundred thirteen **313**

Problem Solving • Applications

3

_____

- - - - -

_____ **take away** _____

_____

- - - - -

4

_____

- - - - -

_____

DIRECTIONS   3. Blair has two marbles. His friend takes one marble from him. Draw to show the subtraction. Write the numbers.   **4.** Write the number that shows how many marbles Blair has now.

HOME ACTIVITY • Show your child a set of four small objects. Have him or her tell how many objects there are. Take one of the objects from the set. Have him or her tell you how many objects there are now.

© Houghton Mifflin Harcourt Publishing Company

Subtraction: Take From

 COMMON CORE STANDARD—K.OA.A.1
Understand addition as putting together and adding to, and understand subtraction as taking apart and taking from.

_____
- - - - -
_____

take away

_____
- - - - -
_____

_____
- - - - -
_____

DIRECTIONS 1. Tell a subtraction word problem about the children. Write the number that shows how many children in all. Write the number that shows how many children are leaving. Write the number that shows how many children are left.

Chapter 6

three hundred fifteen **315**

© Houghton Mifflin Harcourt Publishing Company

Lesson Check (K.OA.A.1)

3 take away 1

_____

- - - - -

_____

Spiral Review (K.CC.B.5, K.OA.A.2)

 + - - - -  - - - -

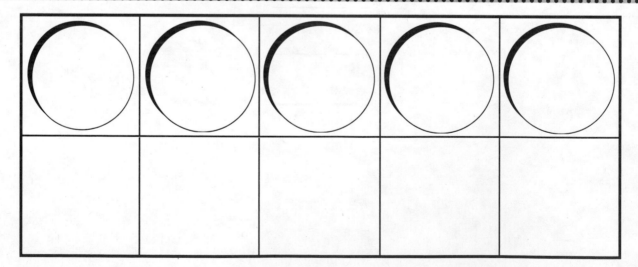

DIRECTIONS 1. Tell a subtraction word problem about the frogs. Write the number that shows how many frogs are left. 2. Tell an addition word problem about the birds. Circle the birds joining the set. Trace and write to complete the addition sentence. 3. How many more counters would you place to model a way to make 8? Draw the counters.

316 three hundred sixteen

© Houghton Mifflin Harcourt Publishing Company

FOR MORE PRACTICE GO TO THE Personal Math Trainer

Name _____

Subtraction: Take Apart

Essential Question How can you show subtraction as taking apart?

 Common Core **Operations and Algebraic Thinking—K.OA.A.1**
MATHEMATICAL PRACTICES
MP2, MP4, MP5

 Listen and Draw Real World

 Hands On

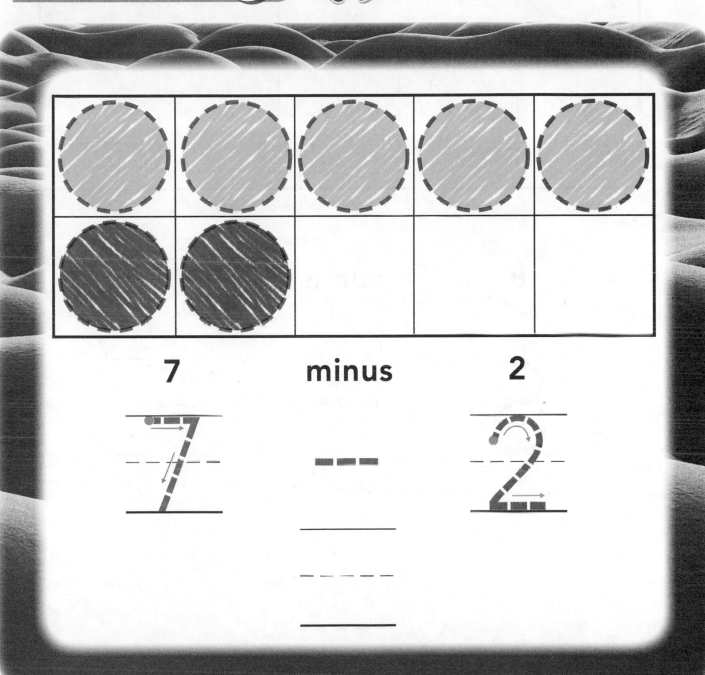

7　　minus　　2

_____

- - - - - - -

_____

DIRECTIONS Listen to the subtraction word problem. Place seven counters in the ten frame as shown. Trace the counters. Trace the number that shows how many in all. Trace the number that shows how many are red. Write the number that shows how many are yellow.

Chapter 6 • Lesson 2

© Houghton Mifflin Harcourt Publishing Company • Image Credits: ©Corbis

❶

<table>
<tr><td></td><td></td><td></td><td></td><td></td></tr>
<tr><td></td><td></td><td></td><td></td><td></td></tr>
</table>

8   minus   1

DIRECTIONS 1. Listen to the subtraction word problem. Place eight counters in the ten frame. Draw and color the counters. Trace the number that shows how many in all. Write the number that shows how many are yellow. Write the number that shows how many are red.

318 three hundred eighteen

© Houghton Mifflin Harcourt Publishing Company

Name _____

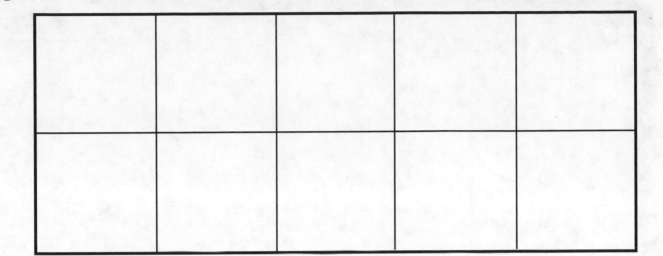

10   minus   4

_____   _____

- - - - - -   ■■■   - - - - - -

_____   _____

_____

- - - - - -

_____

●●●●●●●●●●●●●●●●●●●●●●●●●●●●●●●●●●●●●

DIRECTIONS 2. Listen to the subtraction word problem. Place ten counters in the ten frame. Draw and color the counters. Write the number that shows how many in all. Write the number that shows how many are red. Write the number that shows how many are yellow.

© Houghton Mifflin Harcourt Publishing Company

Chapter 6 • Lesson 2   three hundred nineteen **319**

Problem Solving • Applications — Real World

WRITE Math

3

_____

- - - - -   ▬▬▬   - - - - -

_____

4

_____

- - - - -

_____

DIRECTIONS 3. Juanita has nine apples. One apple is red. The rest of the apples are yellow. Draw the apples. Write the numbers and trace the symbol. **4.** Write the number that shows how many apples are yellow.

HOME ACTIVITY • Show your child a set of seven small objects. Now take away four objects. Have him or her tell a subtraction word problem about the objects.

© Houghton Mifflin Harcourt Publishing Company

Subtraction: Take Apart

Common Core **COMMON CORE STANDARD—K.OA.A.1**
Understand addition as putting together and adding to, and understand subtraction as taking apart and taking from.

1

9   minus   3

_____   _____

— — — —   ▬▬▬   — — — —

_____   _____

_____

— — — —

_____

DIRECTIONS 1. Listen to the subtraction word problem. Jane has nine counters. Three of her counters are red. The rest of her counters are yellow. How many are yellow? Place nine counters in the ten frame. Draw and color the counters. Write the number that shows how many in all. Write the number that shows how many are red. Write the number that shows how many are yellow.

© Houghton Mifflin Harcourt Publishing Company

Chapter 6

three hundred twenty-one **321**

Lesson Check <small>(K.OA.A.1)</small>

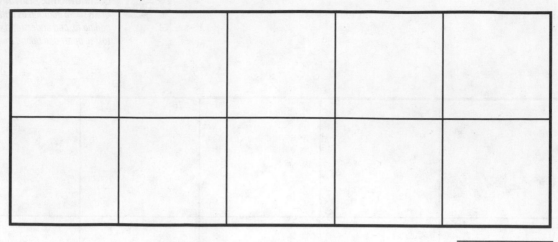

8 − 2

_____

- - - - - - -

_____

Spiral Review <small>(K.CC.C.6)</small>

_____

- - - - - - -

_____

- - - - - - -

_____

DIRECTIONS **1.** Clyde has eight counters. Two of his counters are yellow. The rest of his counters are red. How many are red? Draw and color the counters. Write the number that shows how many are red. **2.** Count the number of leaves in each set. Circle the set that has the greater number of leaves. **3.** Compare the cube trains. Write how many. Circle the number that is greater.

© Houghton Mifflin Harcourt Publishing Company

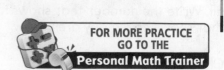

FOR MORE PRACTICE GO TO THE Personal Math Trainer

Name _____

Problem Solving • Act Out Subtraction Problems

Essential Question How can you solve problems using the strategy *act it out*?

 Common Core **Operations and Algebraic Thinking—K.OA.A.1**
Also K.OA.A.2, K.OA.A.5
MATHEMATICAL PRACTICES
MP1, MP2, MP4

 Unlock the Problem Real World

DIRECTIONS Listen to and act out the subtraction word problem. Trace the subtraction sentence. How can you use subtraction to tell how many children are left?

Chapter 6 • Lesson 3

three hundred twenty-three **323**

© Houghton Mifflin Harcourt. Publishing Company

Try Another Problem

  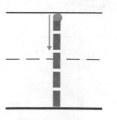 

DIRECTIONS 1. Listen to and act out the subtraction word problem. Trace the numbers and the symbols. Write the number that shows how many children are left.

324 three hundred twenty-four

© Houghton Mifflin Harcourt Publishing Company

Name _____

$$4 - 2 = \underline{\quad}$$

© Houghton Mifflin Harcourt Publishing Company

DIRECTIONS 2. Listen to and act out the subtraction word problem. Trace the numbers and the symbols. Write the number that shows how many children are left.

Chapter 6 • Lesson 3   three hundred twenty-five **325**

On Your Own

WRITE Math

③

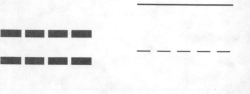

4 − 1 = ___

❀

4 − 3 = ___

DIRECTIONS 3. Tell a subtraction word problem about the kittens. Trace the numbers and the symbols. Write the number that shows how many kittens are left. 4. Draw to show what you know about the subtraction sentence. Write how many are left. Tell a friend a subtraction word problem to match.

 HOME ACTIVITY • Tell your child a short subtraction word problem. Have him or her use objects to act out the word problem.

© Houghton Mifflin Harcourt Publishing Company

Problem Solving • Act Out
Subtraction Problems

 COMMON CORE STANDARD—K.OA.A.1
Understand addition as putting together and adding to, and understand subtraction as taking apart and taking from.

 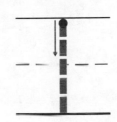 

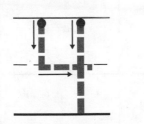  

DIRECTIONS 1. Tell a subtraction word problem about the beavers. Trace the numbers and the symbols. Write the number that shows how many beavers are left. 2. Draw to tell a story about the subtraction sentence. Trace the numbers and the symbols. Write how many are left. Tell a friend about your drawing.

© Houghton Mifflin Harcourt Publishing Company

Chapter 6

1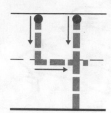

5 -- 4 =

2

3

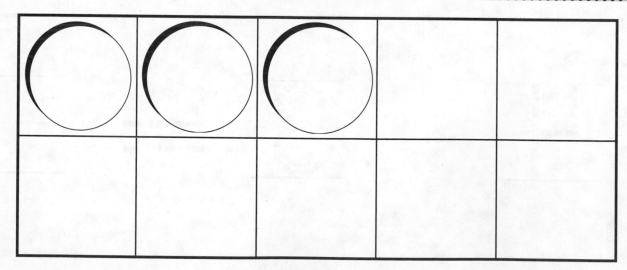

DIRECTIONS 1. Tell a subtraction word problem about the birds. Trace the numbers and the symbols. Write the number that shows how many birds are left. **2.** Count and tell how many bees. Write the number. **3.** How many more counters would you place to model a way to make 7? Draw the counters.

328 three hundred twenty-eight

© Houghton Mifflin Harcourt Publishing Company

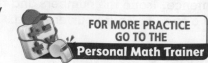
FOR MORE PRACTICE
GO TO THE
Personal Math Trainer

Name _____

Algebra • Model and Draw Subtraction Problems

Essential Question How can you use objects and drawings to solve subtraction word problems?

Common Core **Operations and Algebraic Thinking—K.OA.A.5**
Also K.OA.A.1, K.OA.A.2
MATHEMATICAL PRACTICES
MP1, MP2, MP4

Listen and Draw

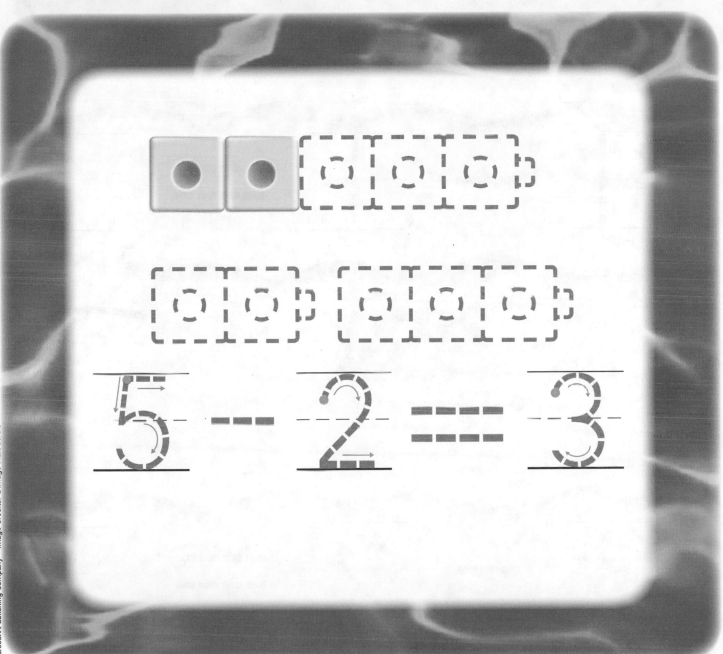

© Houghton Mifflin Harcourt Publishing Company • Image Credits: ©Image Plan/Corbis

DIRECTIONS Model a five-cube train. Two cubes are yellow and the rest are red. Take apart the train to show how many cubes are red. Draw and color the cube trains. Trace the subtraction sentence.

Chapter 6 • Lesson 4

Share and Show

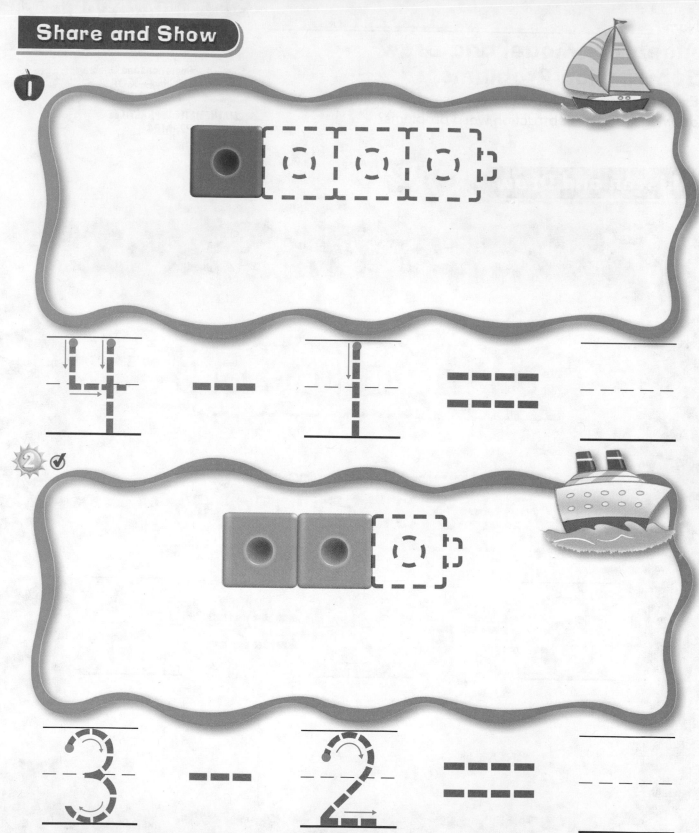

1

4 -- 1 = ___

2 ✓

3 -- 2 = ___

DIRECTIONS 1. Model a four-cube train. One cube is blue and the rest are green. Take apart the train to show how many cubes are green. Draw and color the cube trains. Trace and write to complete the subtraction sentence. 2. Model a three-cube train. Two cubes are orange and the rest are blue. Take apart the train to show how many cubes are blue. Draw and color the cube trains. Trace and write to complete the subtraction sentence.

330 three hundred thirty

© Houghton Mifflin Harcourt Publishing Company

Name _____

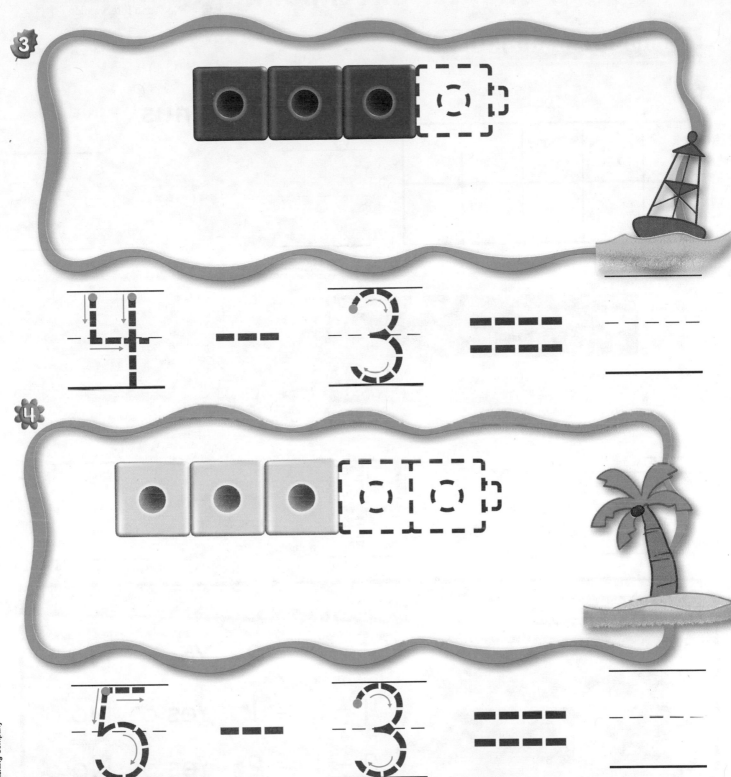

3

4 − 3 = _____

4

5 − 3 = _____

DIRECTIONS 3. Model a four-cube train. Three cubes are red and the rest are blue. Take apart the train to show how many cubes are blue. Draw and color the cube trains. Trace and write to complete the subtraction sentence. 4. Model a five-cube train. Three cubes are yellow and the rest are green. Take apart the train to show how many cubes are green. Draw and color the cube trains. Trace and write to complete the subtraction sentence.

HOME ACTIVITY • Show your child two small objects. Take apart the set of objects. Have him or her tell a word problem to match the subtraction.

© Houghton Mifflin Harcourt Publishing Company

Personal Math Trainer
Online Assessment
and Intervention

Concepts and Skills

6   **minus**   **1**

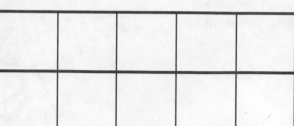

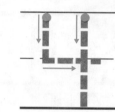 

3 **THINK** SMARTER

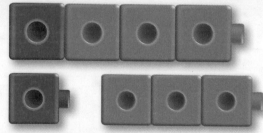

$4 - 2 = 2$ Yes ○   No ○

$4 - 3 = 1$ Yes ○   No ○

$3 - 1 = 2$ Yes ○   No ○

DIRECTIONS **1.** Choi has 6 counters. One of his counters is yellow. The rest are red. Draw and color the six counters in the ten frame. Write the number that shows how many in all. Write the number that shows how many are yellow. (K.OA.A.1) **2.** Model a five-cube train. Four cubes are blue and the rest are orange. Take apart the cube train to show how many are orange. Draw and color the cube trains. Trace and write to complete the subtraction sentence. (K.OA.A.5) **3.** Choose Yes or No. Does the subtraction sentence match the model? (K.OA.A.5)

© Houghton Mifflin Harcourt Publishing Company

Algebra • Model and Draw Subtraction Problems

COMMON CORE STANDARD—K.OA.A.5
Understand addition as putting together and adding to, and understand subtraction as taking apart and taking from.

  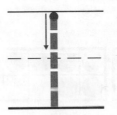 

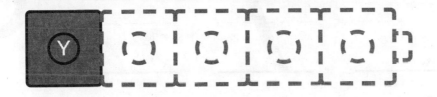

DIRECTIONS I. Model a three-cube train. Two cubes are red and the rest are blue. Take apart the cube train to show how many cubes are blue. Draw and color the cube trains. Trace and write to complete the subtraction sentence. 2. Model a five-cube train. One cube is yellow and the rest are green. Take apart the train to show how many cubes are green. Draw and color the cube trains. Trace and write to complete the subtraction sentence.

© Houghton Mifflin Harcourt Publishing Company

Lesson Check (K.OA.A.5)

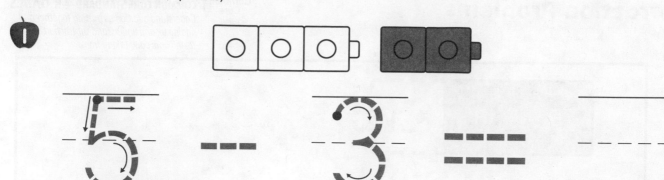

Spiral Review (K.CC.A.2, K.OA.A.3)

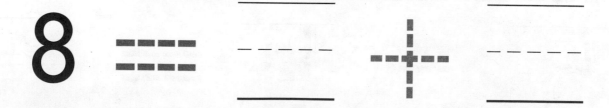

5   **7**   **9**

3

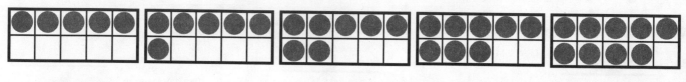

8

DIRECTIONS 1. Ellie makes the cube train shown. She takes the cube train apart to show how many cubes are gray. Trace and write to show the subtraction sentence for Ellie's cube train. 2. Count the dots in the ten frames. Begin with 5. Write the numbers in order as you count forward. 3. Complete the addition sentence to show the numbers that match the cube train.

334 three hundred thirty-four

© Houghton Mifflin Harcourt Publishing Company

**FOR MORE PRACTICE
GO TO THE
Personal Math Trainer**

Name _____

Algebra • Write Subtraction Sentences

Essential Question How can you solve subtraction word problems and complete the equation?

Common Core
Operations and Algebraic Thinking—K.OA.A.5
Also K.OA.A.1, K.OA.A.2
MATHEMATICAL PRACTICES
MP1, MP2

Listen and Draw

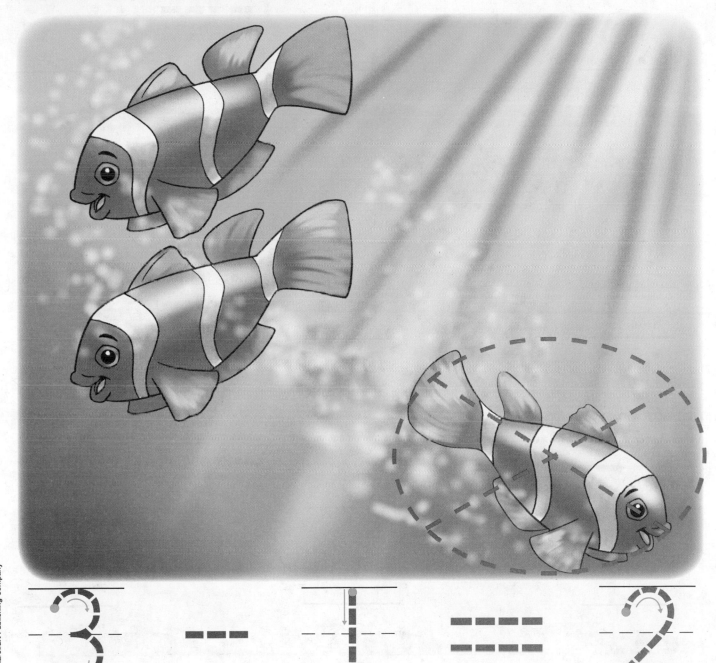

$$3 - 1 = 2$$

DIRECTIONS There are three fish. Some fish swim away. Now there are two fish. Trace the circle and X to show the fish swimming away. Trace the subtraction sentence.

© Houghton Mifflin Harcourt Publishing Company

Share and Show

1.

$$5 - 2 = 3$$

2.

$$4 - \_\_\_ = 3$$

3. ✓

$$4 - \_\_\_ = 1$$

DIRECTIONS 1. Listen to the subtraction word problem. Trace the circle and X to show how many are being taken from the set. Trace to complete the subtraction sentence. 2–3. Listen to the subtraction word problem. How many are being taken from the set? Circle and mark an X to show how many are being taken from the set. Trace and write to complete the subtraction sentence.

© Houghton Mifflin Harcourt Publishing Company

Name _____

4

5 -- ___ == 2

5

3 -- ___ == 1

6

5 -- ___ == 1

DIRECTIONS 4–6. Listen to the subtraction word problem. How many are being taken from the set? Circle and mark an X to show how many are being taken from the set. Trace and write to complete the subtraction sentence.

© Houghton Mifflin Harcourt Publishing Company

Problem Solving • Applications Real World

WRITE Math

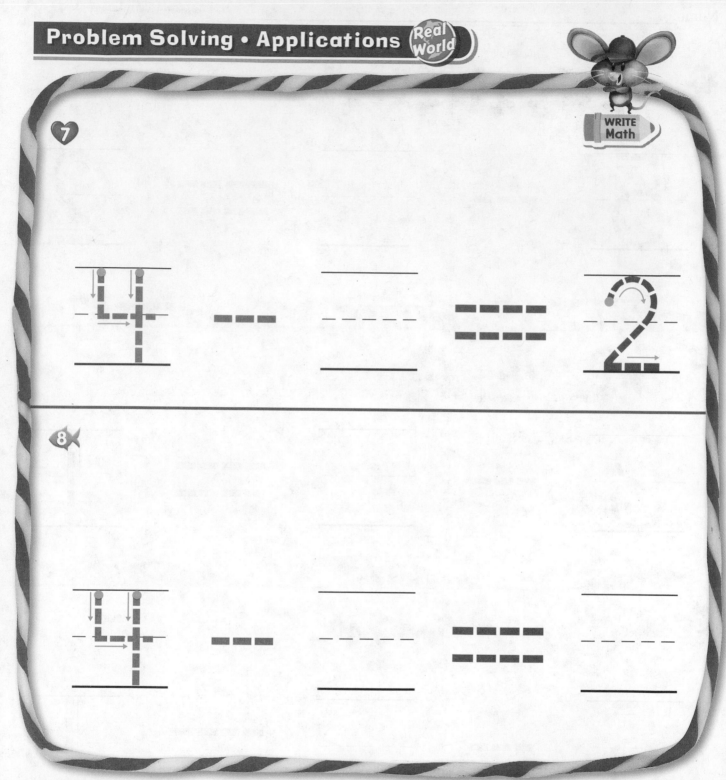

7

$4 - \underline{} = 2$

8

$4 - \underline{} = \underline{}$

DIRECTIONS 7. Kristen has four flowers. She gives her friend some flowers. Now Kristen has two flowers. How many did Kristen give her friend? Draw to solve the problem. Complete the subtraction sentence. 8. Tell a different subtraction word problem about the flowers. Draw to solve the problem. Tell a friend about your drawing. Complete the subtraction sentence.

HOME ACTIVITY • Have your child draw a set of five or fewer balloons. Have him or her circle and mark an X on some balloons to show that they have popped. Then have your child tell a word problem to match the subtraction.

338 three hundred thirty-eight

© Houghton Mifflin Harcourt Publishing Company

Name _____

Algebra • Write Subtraction Sentences

 COMMON CORE STANDARD—K.OA.A.5
Understand addition as putting together and adding to, and understand subtraction as taking apart and taking from.

 1

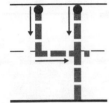 —  =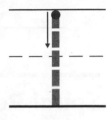

 2

 —  =

 3

 —    =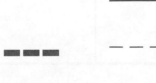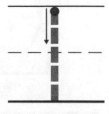

DIRECTIONS 1–3. Listen to the subtraction word problem about the animals. There are ____ ____. Some are taken from the set. Now there are ____. How many are taken from the set? Circle and mark an X to show how many are being taken from the set. Trace and write to complete the subtraction sentence.

Chapter 6   three hundred thirty-nine **339**

© Houghton Mifflin Harcourt Publishing Company

Lesson Check (K.OA.A.5)

1

 — — — — — —

Spiral Review (K.CC.B.5, K.CC.C.6)

2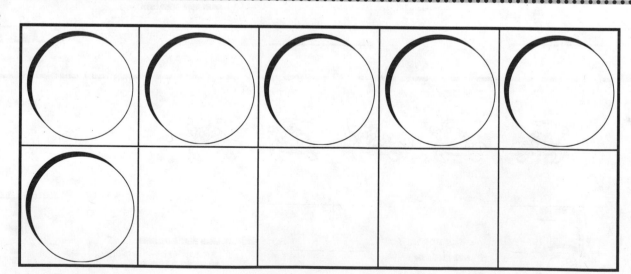

3

DIRECTIONS 1. Trace and write to show the subtraction sentence for the set. 2. Count the number of counters in each set. Circle the set that has the greater number of counters. 3. How many more counters would you place to model a way to make 9? Draw the counters.

340 three hundred forty

© Houghton Mifflin Harcourt Publishing Company

FOR MORE PRACTICE GO TO THE
Personal Math Trainer

Name _____

Algebra • Write More Subtraction Sentences

Essential Question How can you solve subtraction word problems and complete the equation?

Common Core — **Operations and Algebraic Thinking—K.OA.A.2**
Also K.OA.A.1
MATHEMATICAL PRACTICES
MP1, MP2

Listen and Draw *Real World*

© Houghton Mifflin Harcourt Publishing Company

DIRECTIONS There are six birds. A bird flies away. Trace the circle and X around that bird. How many birds are left? Trace the subtraction sentence.

Chapter 6 • Lesson 6

three hundred forty-one **341**

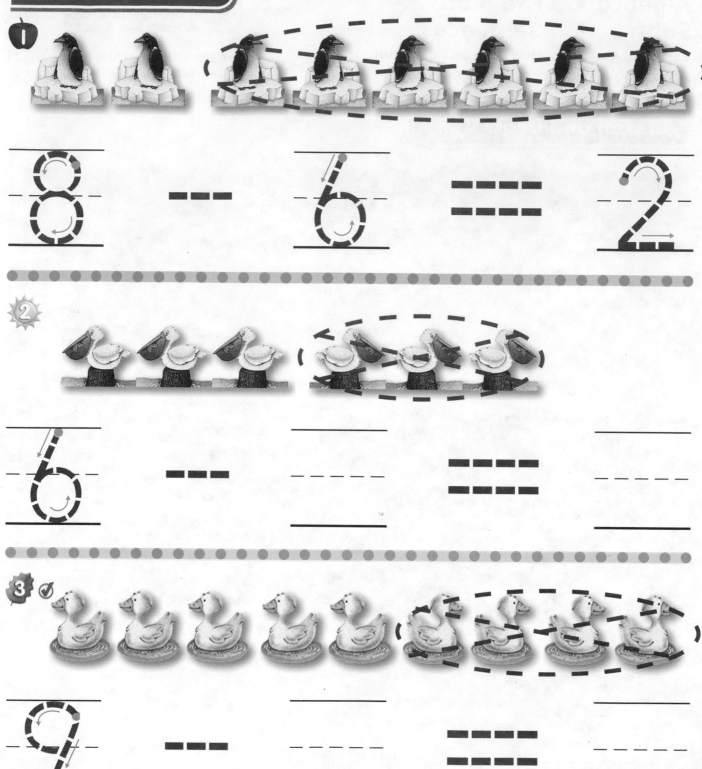

DIRECTIONS Listen to the subtraction word problem. **1.** How many birds are taken from the set? Trace the circle and X. How many birds are left? Trace the subtraction sentence. **2–3.** How many birds are taken from the set? Trace the circle and X. How many birds are left? Trace and write to complete the subtraction sentence.

© Houghton Mifflin Harcourt Publishing Company

© Houghton Mifflin Harcourt Publishing Company

Name _____

4

6   —   _____   =   _____

5

9   —   _____   =   _____

6

8   —   _____   =   _____

DIRECTIONS 4–6. Listen to the subtraction word problem. How many birds are taken from the set? Trace the circle and X. How many birds are left? Trace and write to complete the subtraction sentence.

Problem Solving • Applications Real World

7

8 — — — — — — — — — — — — — —

DIRECTIONS 7. Complete the subtraction sentence. Draw a picture of real objects to show what you know about this subtraction sentence. Tell a friend about your drawing.

HOME ACTIVITY • Tell your child you have ten small objects in your hand. Tell him or her that you are taking two objects from the set. Ask him or her to tell you how many objects are in your hand now.

© Houghton Mifflin Harcourt Publishing Company

Algebra • Write More Subtraction Sentences

 COMMON CORE STANDARD—K.OA.A.2
Understand addition as putting together and adding to, and understand subtraction as taking apart and taking from.

 ❶

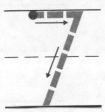

7 ___ ▭ ___ ___ ___

───────────────────────

② ❷

9 ___ ▭ ___ ___ ___

───────────────────────

③ ❸

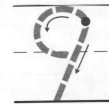

6 ___ ▭ ___ ___ ___

───────────────────────

© Houghton Mifflin Harcourt Publishing Company

DIRECTIONS 1–3. Listen to a subtraction word problem about the birds. There are seven birds. _____ birds are taken from the set. Now there are _____ birds. How many birds are taken from the set? How many birds are there now? Trace and write to complete the subtraction sentence.

1

 --- === -----

2

_____

- - - - - - - - - -

_____

3

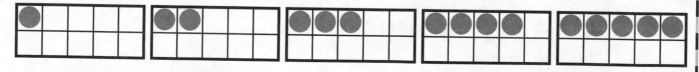

1   2      4   - - - - - -

DIRECTIONS   1. Trace and write to show the subtraction sentence for the buses.   **2.** How many lunch boxes are there? Write the number.   **3.** Count the dots in the ten frames. Begin with 1. Write the numbers in order as you count forward.

© Houghton Mifflin Harcourt Publishing Company

FOR MORE PRACTICE GO TO THE Personal Math Trainer

Name _____

Algebra • Addition and Subtraction

Essential Question How can you solve word problems using addition and subtraction?

Common Core Operations and Algebraic Thinking—K.OA.A.2
Also K.OA.A.1

MATHEMATICAL PRACTICES
MP2, MP5, MP8

Listen and Draw

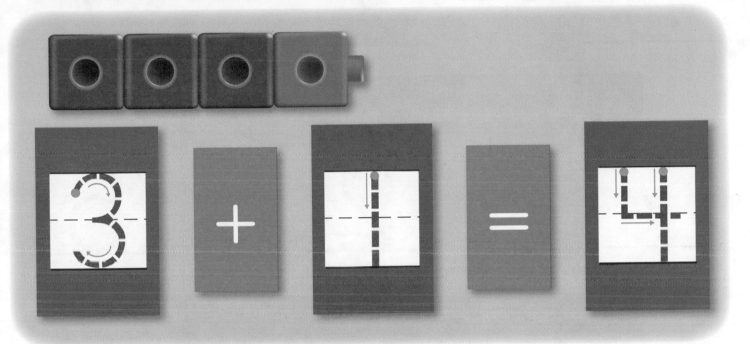

DIRECTIONS Listen to the addition and subtraction word problems. Use cubes and Number and Symbol Tiles as shown to match the word problems. Trace to complete the number sentences.

Chapter 6 • Lesson 7

© Houghton Mifflin Harcourt Publishing Company

Share and Show

1

$$5 + 2 = 7$$

$$7 - 2 = 5$$

2

___ + ___ = ___

___ - ___ = ___

DIRECTIONS Tell addition and subtraction word problems. Use cubes to add and to subtract. **1.** Trace the number sentences. **2.** Complete the number sentences.

348 three hundred forty-eight

© Houghton Mifflin Harcourt Publishing Company

Name _____

3

‑ ‑ ‑ ‑ $+$ ‑ ‑ ‑ ‑ $=$ ‑ ‑ ‑ ‑
___   ___   ___   ___

‑ ‑ ‑ ‑ $-$ ‑ ‑ ‑ ‑ $=$ ‑ ‑ ‑ ‑
___   ___   ___   ___

4

‑ ‑ ‑ ‑ $+$ ‑ ‑ ‑ ‑ $=$ ‑ ‑ ‑ ‑
___   ___   ___   ___

‑ ‑ ‑ ‑ $-$ ‑ ‑ ‑ ‑ $=$ ‑ ‑ ‑ ‑
___   ___   ___   ___

DIRECTIONS 3–4. Tell addition and subtraction word problems. Use cubes to add and subtract. Complete the number sentences.

© Houghton Mifflin Harcourt Publishing Company

Problem Solving • Applications Real World

WRITE Math

6 + 3 = 9

5

_____ _____ _____

– – – – – ▬▬▬ – – – – – ▬▬▬▬ – – – – –
▬▬▬▬

_____ _____ _____

6

_____ _____ _____

– – – – – ▬▬▬ – – – – – ▬▬▬▬ – – – – –
▬▬▬▬

_____ _____ _____

DIRECTIONS Look at the addition sentence at the top of the page. **5–6.** Tell a related subtraction word problem. Complete the subtraction sentence.

HOME ACTIVITY • Ask your child to use objects to model a simple addition problem. Then have him or her explain how to make it into a subtraction problem.

© Houghton Mifflin Harcourt Publishing Company

Algebra • Addition and Subtraction

 COMMON CORE STANDARD—K.OA.A.2
*Understand addition as putting together and
adding to, and understand subtraction as
taking apart and taking from.*

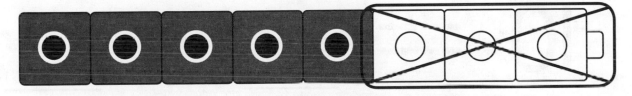

DIRECTIONS 1–2. Tell an addition or subtraction word
problem. Use cubes to add or subtract. Complete the
number sentence.

© Houghton Mifflin Harcourt Publishing Company

Lesson Check (K.OA.A.2)

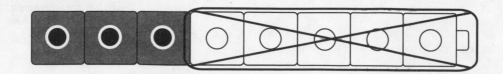

_____   _____   _____

– – – –   ▬ ▬ ▬   – – – –   ▬ ▬ ▬   – – – –
                              ▬ ▬ ▬

_____   _____   _____

Spiral Review (K.CC.C.7, K.OA.A.3)

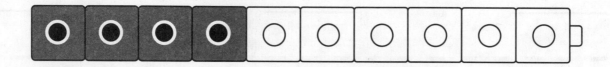

10   ▬ ▬ ▬   _____ ➕ _____
      ▬ ▬ ▬   – – –   – – –
             _____   _____

8   9

DIRECTIONS 1. Tell a subtraction word problem. Use cubes to subtract. Complete the number sentence. 2. Complete the addition sentence to show the numbers that match the cube train. 3. Compare the numbers. Circle the number that is greater.

FOR MORE PRACTICE
GO TO THE
Personal Math Trainer

© Houghton Mifflin Harcourt Publishing Company

✓ Chapter 6 Review/Test

1

4 take away _____

_____

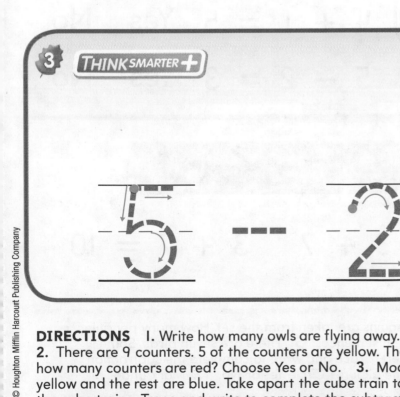

2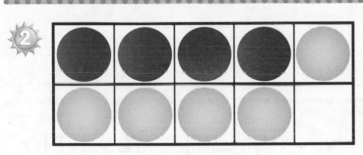

$9 - 1$ ○ Yes ○ No

$9 - 5$ ○ Yes ○ No

$8 - 3$ ○ Yes ○ No

Personal Math Trainer

3 THINK SMARTER ➕

$5 - 2 = \text{\_\_\_\_}$

DIRECTIONS 1. Write how many owls are flying away. Write how many owls are left.
2. There are 9 counters. 5 of the counters are yellow. The rest are red. Which answers show how many counters are red? Choose Yes or No. **3.** Model a five-cube train. Two cubes are yellow and the rest are blue. Take apart the cube train to show how many are blue. Draw the cube trains. Trace and write to complete the subtraction sentence.

© Houghton Mifflin Harcourt Publishing Company

GO DIGITAL Assessment Options **Chapter Test**

4

4 -- 2 == ===

5

7 -- === -----

6

5 − 4 = 1	Yes	No
4 + 1 = 5	Yes	No
5 − 2 = 3	Yes	No

7

9 = 3 + 6   10 = 3 + 7   3 + 7 = 10

     ○   ○   ○

DIRECTIONS **4.** There are 4 penguins. Two penguins are taken from the set. How many penguins are left? Trace and write to complete the subtraction sentence. **5.** There are seven birds. Some birds are taken from the set. How many birds are left? Trace and write to complete the subtraction sentence. **6.** Does the number sentence match the picture? Circle Yes or No. **7.** Mark under all the number sentences that match the cubes.

© Houghton Mifflin Harcourt Publishing Company

4 -- 3 == _____

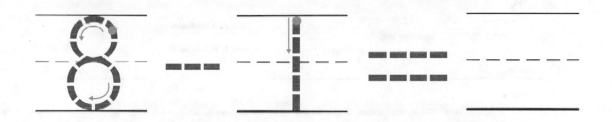

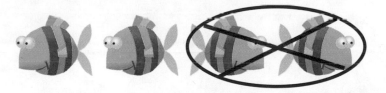

8 -- 1 == _____

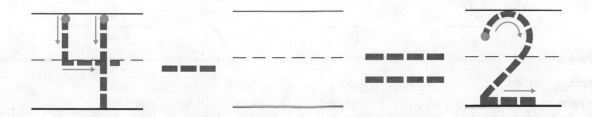

4 -- _____ == _____ 2

DIRECTIONS **8.** Model a four-cube train. Three cubes are red and the rest are blue. Take apart the train to show how many cubes are blue. Draw the cube trains. Complete the subtraction sentence. **9–10.** Complete the subtraction sentence to match the picture.

© Houghton Mifflin Harcourt Publishing Company

 THINK SMARTER +

_____ __ _____ == **0**

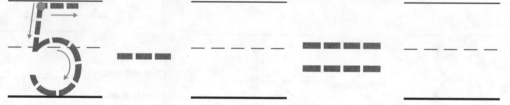

5 -- _____ === _____

- - - - - - - - - - - - - - - - - - - -

6 -- _____ === **4**

- - - - - - - - - - - - - - - - - - - -

DIRECTIONS **11.** There were some apples on a tree. Some were taken away. Now there are zero apples left. Draw to show how many apples there could have been to start. Cross out apples to show how many were taken away. Complete the subtraction sentence. **12.** There are five birds. Some birds are taken from the set. How many birds are left? Trace and write to complete the subtraction sentence. **13.** Erica has 6 balloons. She gives some of her balloons to a friend. Now Erica has 4 balloons. How many did Erica give to her friend? Draw to solve the problem. Complete the subtraction sentence.

© Houghton Mifflin Harcourt Publishing Company